ENERGY SECTOR STANDARD OF THE PEOPLE'S REPUBLIC OF CHINA

中华人民共和国能源行业标准

Specification for Preparation of Planning Reports for Offshore Wind Power Projects

海上风电场工程规划报告编制规程

NB/T 31108-2017

Chief Development Department: China Renewable Energy Engineering Institute
Approval Department: National Energy Administration of the People's Republic of China
Implementation Date: August 1, 2017

China Water & Power Press

Beijing 2024

All rights reserved. No part of this publication may be reproduced, stored in a retrieval system, or transmitted in any form or by any means—electronic, mechanical, photocopying, recording or otherwise, without prior written permission of the publisher.

图书在版编目（CIP）数据

海上风电场工程规划报告编制规程：NB/T 31108 -2017 = Specification for Preparation of Planning Reports for Offshore Wind Power Projects (NB/T 31108-2017)：英文 / 国家能源局发布. -- 北京：中国水利水电出版社, 2024. 10. -- ISBN 978-7-5226 -2767-0

I. TM614-65

中国国家版本馆CIP数据核字第2024VU0798号

ENERGY SECTOR STANDARD
OF THE PEOPLE'S REPUBLIC OF CHINA
中华人民共和国能源行业标准

Specification for Preparation of Planning Reports
for Offshore Wind Power Projects
海上风电场工程规划报告编制规程
NB/T 31108-2017
（英文版）

Issued by National Energy Administration of the People's Republic of China
国家能源局　发布
Translation organized by China Renewable Energy Engineering Institute
水电水利规划设计总院　组织翻译
Published by China Water & Power Press
中国水利水电出版社　出版发行
　　Tel: (+ 86 10) 68545888　68545874
　　sales@mwr.gov.cn
　　Account name: China Water & Power Press
　　Address: No.1, Yuyuantan Nanlu, Haidian District, Beijing 100038, China
　　http://www.waterpub.com.cn
中国水利水电出版社微机排版中心　排版
北京中献拓方科技发展有限公司　印刷
184mm×260mm　16开本　2.25印张　71千字
2024年10月第1版　2024年10月第1次印刷

Price（定价）：￥320.00

Introduction

This English version is one of China's energy sector standard series in English. Its translation was organized by China Renewable Energy Engineering Institute authorized by National Energy Administration of the People's Republic of China in compliance with relevant procedures and stipulations. This English version was issued by National Energy Administration of the People's Republic of China in Announcement [2022] No. 4 dated May 13, 2022.

This version was translated from the Chinese Standard NB/T 31108-2017, *Specification for Preparation of Planning Reports for Offshore Wind Power Projects*, published by China Electric Power Press. The copyright is reserved by National Energy Administration of the People's Republic of China. In the event of any discrepancy in the implementation, the Chinese version shall prevail.

Many thanks go to the staff from the relevant standard development organizations and those who have provided generous assistance in the translation and review process.

For further improvement of the English version, any comments and suggestions are welcome and should be addressed to:

China Renewable Energy Engineering Institute
No. 2 Beixiaojie, Liupukang, Xicheng District, Beijing 100120, China
Website: www.creei.cn

Translating organization:

POWERCHINA Zhongnan Engineering Corporation Limited

Translating staff:

YANG Hong   LIU Xiaofen   CHEN Lei   LI Qian

Review panel members:

QIE Chunsheng	Senior English Translator
YAN Wenjun	Army Academy of Armored Forces, PLA
QI Wen	POWERCHINA Beijing Engineering Corporation Limited
ZHANG Ming	Tsinghua University
LI Kejia	POWERCHINA Northwest Engineering Corporation Limited
LI Yu	POWERCHINA Huadong Engineering Corporation

	Limited
YU Dingkun	POWERCHINA Chengdu Engineering Corporation Limited
ZHONG Yao	POWERCHINA Zhongnan Engineering Corporation Limited
LI Shisheng	China Renewable Energy Engineering Institute

National Energy Administration of the People's Republic of China

翻译出版说明

本译本为国家能源局委托水电水利规划设计总院按照有关程序和规定，统一组织翻译的能源行业标准英文版系列译本之一。2022年5月13日，国家能源局以2022年第4号公告予以公布。

本译本是根据中国电力出版社出版的《海上风电场工程规划报告编制规程》NB/T 31108—2017翻译的，著作权归国家能源局所有。在使用过程中，如出现异议，以中文版为准。

本译本在翻译和审核过程中，本标准编制单位及编制组有关成员给予了积极协助。

为不断提高本译本的质量，欢迎使用者提出意见和建议，并反馈给水电水利规划设计总院。

地址：北京市西城区六铺炕北小街2号
邮编：100120
网址：www.creei.cn

本译本翻译单位：中国电建集团中南勘测设计研究院有限公司
本译本翻译人员：杨　虹　刘小芬　陈　蕾　李　倩
本译本审核人员：

郄春生　英语高级翻译
闫文军　中国人民解放军陆军装甲兵学院
齐　文　中国电建集团北京勘测设计研究院有限公司
张　明　清华大学
李可佳　中国电建集团西北勘测设计研究院有限公司
李　瑜　中国电建集团华东勘测设计研究院有限公司
俞定坤　中国电建集团成都勘测设计研究院有限公司
钟　耀　中国电建集团成都勘测设计研究院有限公司
李仕胜　水电水利规划设计总院

国家能源局

Announcement of National Energy Administration of the People's Republic of China [2017] No. 6

According to the requirements of Document GNJKJ [2009] No. 52, "Notice on Releasing the Energy Sector Standardization Administration Regulations (*tentative*) and detailed implementation rules issued by National Energy Administration of the People's Republic of China", 159 sector standards such as *Shale Gas Reservoir Stimulation—Part 2: Technical Specification for Factory Fracturing Operation*, including 34 energy standards (NB), 39 electric power standards (DL), and 86 petroleum standards (SY), are issued by National Energy Administration of the People's Republic of China after due review and approval.

The electric power standards are published by China Electric Power Press or China Planning Press; coal standards by China Coal Industry Publishing House; petroleum, natural gas and shale gas standards by Petroleum Industry Press; and boiler and pressure vessel standards by Xinhua Publishing House.

Attachment: Directory of Sector Standards

National Energy Administration of the People's Republic of China

March 28, 2017

Attachment:

Directory of Sector Standards

Serial number	Standard No.	Title	Replaced standard No.	Adopted international standard No.	Approval date	Implementation date
...						
11	NB/T 31108-2017	Specification for Preparation of Planning Reports for Offshore Wind Power Projects			2017-03-28	2017-08-01
...						

Foreword

According to the requirements of Document GNKJ [2012] No. 83 issued by National Energy Administration of the People's Republic of China, "Notice on Releasing the Development and Revision Plan of the First Batch of Energy Sector Standards in 2012", and after extensive investigation and research, summarization of the practical experience in offshore wind farm project development at home and abroad, and wide solicitation of opinions, the drafting group has prepared this specification.

The main technical contents of this specification include: general, planning principles and basis, siting, project construction conditions, planned installed capacity, preliminary power system interconnection scheme, planning scheme of typical wind farm, preliminary environmental impact assessment, rough investment estimation, preliminary analysis of benefits, planning objectives and development priorities, and conclusions and suggestions.

National Energy Administration of the People's Republic of China is in charge of the administration of this specification. China Renewable Energy Engineering Institute has proposed this specification and is responsible for its routine management. Sub-committee on Planning and Design of Wind Power Project of Energy Sector Standardization Technical Committee on Wind Power is responsible for the explanation of specific technical contents. Comments and suggestions in the implementation of this specification should be addressed to:

China Renewable Energy Engineering Institute
No. 2 Beixiaojie, Liupukang, Xicheng District, Beijing 100120, China

Chief development organization:

POWERCHINA Zhongnan Engineering Corporation Limited

Participating development organizations:

POWERCHINA Huadong Engineering Corporation Limited

POWERCHINA Northwest Engineering Corporation Limited

POWERCHINA Beijing Engineering Corporation Limited

Chief drafting staff:

LIU Xiaosong	FANG Tao	FU Liangming	WU Jinhua
ZHONG Yao	XIE Yuefei	YAO Xiyu	HUANG Chunfang
LIU Wei	LIU Guopin	PENG Tiankui	LIU Congzhu
ZHOU Kai	WANG Tao	YE Fengyi	CAO Yuanyuan

Review panel members:

XIE Hongwen	ZHAO Shengxiao	DONG Delan	LI Jianying
WEI Xijian	SHI Lei	MI Youwan	LI Fagui
YANG Baoliu	ZENG Zhenliang	WANG Xiaoying	HAI Shen
NING Hongtao	WEI Hui	ZHOU Bing	WEN Peng
YE Rui	ZOU Hui	CHI Hongming	LI Shisheng

Contents

1	**General Provisions**	1
2	**Terms**	2
3	**Basic Requirements**	3
4	**General**	5
5	**Planning Principles and Basis**	6
6	**Siting**	7
6.1	Principles and Requirements	7
6.2	Outcomes	7
7	**Project Construction Conditions**	8
7.1	Wind Energy Resources	8
7.2	Marine Hydrology	8
7.3	Engineering Geology	8
7.4	Transportation and Construction	9
8	**Planned Installed Capacity**	10
9	**Preliminary Power System Interconnection Scheme**	11
10	**Planning Scheme of Typical Wind Farm**	12
11	**Preliminary Environmental Impact Assessment**	13
12	**Rough Investment Estimation**	14
13	**Preliminary Analysis of Benefits**	15
14	**Planning Objectives and Development Priorities**	16
15	**Conclusions and Suggestions**	17
Appendix A	**Contents of Planning Report for Offshore Wind Power Project**	18
Explanation of Wording in This Specification		21

1 General Provisions

1.0.1 This specification is formulated with a view to standardizing the contents, depth and requirements for the preparation of planning reports for offshore wind power projects.

1.0.2 This specification is applicable to the preparation of planning reports for offshore wind power projects except deep-sea wind farms.

1.0.3 In addition to this specification, the preparation of planning reports for offshore wind power projects shall comply with other current relevant standards of China.

2 Terms

2.0.1 coefficient of installed capacity

installed capacity (MW) per square kilometer of planned sea area

3 Basic Requirements

3.0.1 The planning of offshore wind power projects shall follow the principles of environmental protection, overall coordination, multipurpose utilization, and rational exploitation; comply with the national economic development plan and energy development plan; and be coordinated with the marine functional zoning, sea islands protection plan, coastal protection and utilization plan, marine environment protection plan, shipping route plan, etc.

3.0.2 The planning of offshore wind power projects shall be based on the investigation of the basic conditions of the offshore area and the data collection, arrangement and analysis. The planning scheme shall be demonstrated and optimized in the principles of coordination, efficiency and cost-effectiveness.

3.0.3 The planning of offshore wind power projects shall rationally determine the planning target years, and should consider the planning target years for the short term, medium term and long term with emphasis on the short term. The planning target years shall be determined in agreement with the medium- and long-term national development plans.

3.0.4 The following basic data shall be collected at the planning stage:

1 The current socio-economic status and development plan of the region where the planned wind farm is located.

2 The energy resources and development situation and the current status and development plan of the electric power system in the region where the planned wind farm is located.

3 The anemometry data of the planned offshore area.

4 The data of the planned offshore area and the long series reference station nearby, including:

   1) The location, coordinates, altitude, and historic relocation of the reference station; the model and installation height of observation instruments and the variation of the surrounding buildings; or the results of data reanalysis and adaptability analysis.

   2) Statistic data on air temperature, atmospheric pressure, humidity, precipitation, thunderstorm, hailstone, solar radiation, ice, snow, fog, etc.

   3) Monthly average wind speed, maximum wind speed, extreme wind speed, and average annual wind direction frequency.

   4) Data on tropical cyclones.

5   Marine hydrological data of the planned offshore area or the nearby marine hydrometric station, including tide, ocean current, wave, sea ice, and sediment.

6   The sea chart of the planned offshore area with a scale no less than 1 : 250 000 and the topographic map of the nearby onshore area with a scale no less than 1 : 50 000.

7   The engineering geological data of the planned offshore area and the nearby onshore area.

8   The marine economic development plan, marine function zoning, coastal development plan, seaport development plan, marine environment protection plan, marine island protection plan, shipping route plan, and marine traffic flow monitoring chart of the last year of the planned offshore area.

9   The current distribution and planning of submarine cables, optical cables, gas and oil pipelines, buildings (structures) in the planned offshore area and its vicinity.

10 The marine nature reserves, important tourism areas, key fishery and aquaculture areas, special marine reserves, typical marine ecosystems, natural and historical relics reserves, identified important mineral resources, etc. in the planned offshore area and its vicinity.

11 Current financial and tax policies.

12 Administrative division of the planned offshore area.

3.0.5 The following basic data should be collected at the planning stage:

1   Basic information about the existing meteorological masts around the planned offshore area and available wind energy observation data.

2   Authority division between oceanic administration and maritime administration for the planned offshore area.

3.0.6 Opinions of related authorities of the military sea area and airspace as well as civil aviation in the planned offshore area and its vicinity shall be consulted in the process of preparing the plan.

3.0.7 The contents of the planning report of an offshore wind power project should comply with Appendix A of this specification.

4　General

4.0.1　The background of planning shall be briefly stated.

4.0.2　The physical geography, meteorology, socio-economy, transportation, energy resources, electric power system, marine environment and resources, and marine function zoning of the region where the planned offshore wind power project is located as well as an overview of the preliminary works shall be briefly described.

4.0.3　The main contents and outcomes of planning shall be briefly stated.

5 Planning Principles and Basis

5.0.1 The guidelines and principles for the offshore wind power project planning shall be briefly described.

5.0.2 The basis for the offshore wind power project planning shall be listed, and the planning scope and target year shall be stated.

6 Siting

6.1 Principles and Requirements

6.1.1 The offshore wind energy resource distribution in the planned area shall be briefly described based on the wind energy resource survey results and the wind energy resource measurement and evaluation, and the offshore area exploitable for wind energy resource shall be analyzed and identified.

6.1.2 The siting of the offshore wind farm shall be coordinated with the marine function zoning, coastline development plan, seaport development plan, marine environment protection plan, sea islands protection plan, and shipping route plan, and shall meet the following requirements:

1. Avoid sensitive offshore areas such as marine nature reserves, special marine reserves, and historic relics reserves.

2. Avoid port areas, anchorages, navigation channels, dense navigation areas, and navigation routes published by the competent authorities.

3. Avoid restricted areas such as military offshore areas and airspace and civil aviation areas.

4. Comply with stipulations relating to marine natural environment protection.

5. Be coordinated with the submarine pipelines, offshore platforms, important tourism areas, and other special-purpose offshore areas.

6. Give consideration to the administrative division of the offshore area.

6.2 Outcomes

6.2.1 The planning area for the offshore wind farm shall be proposed in accordance with the siting principles and requirements, taking into account the constraints to the site planning and the project construction conditions.

6.2.2 The range of each offshore wind farm site shall be proposed, taking into account the construction conditions, grid connection, navigation safety, scale benefits, and mutual influence among wind farms. The area of each wind farm shall be put forward, and the layout of the planned wind farms and the site boundary map shall be plotted.

7 Project Construction Conditions

7.1 Wind Energy Resources

7.1.1 The regional climatic characteristics shall be briefly described.

7.1.2 Based on the long series data from the reference station of the planned wind farm and the outcomes of the relevant wind energy resource evaluation, the inter-annual and intra-annual variation patterns of the wind regime in the region shall be analyzed, the wind energy resources shall be evaluated, and the inter-annual and intra-annual wind speed variation histograms and wind direction rose diagrams of the reference station shall be plotted.

7.1.3 The representativeness of the measured data of the planned wind farm site shall be analyzed according to the measured wind energy resource data of the planned wind farm, the location of the meteorological station, and the long series data from the reference station.

7.1.4 The wind characteristics of the planned wind farm shall be analyzed, and the wind parameters of the wind farms shall be proposed, including the annual average wind speed, wind power density, annual and daily distributions of wind speed and wind power density, distribution of wind direction and wind energy direction, wind shear, turbulence intensity, and the maximum 10-min mean wind speed with a return period of 50 years. The wind energy resource at the wind farm site shall be evaluated, and the annual and daily variation curves of wind speed versus wind power density of one whole year, wind direction and wind energy rose diagrams, wind speed and wind energy-frequency distribution histograms of the planned wind farm shall be plotted.

7.1.5 The impact of adverse climatic conditions such as tropical cyclones on the planned wind farm shall be analyzed based on the related data.

7.2 Marine Hydrology

7.2.1 The marine hydrological characteristics of the sea area where the wind farm is located shall be preliminarily analyzed.

7.2.2 The marine hydrological characteristics shall include parameters such as tide, current, wave, ice, and sediment.

7.3 Engineering Geology

7.3.1 The engineering geology of the planned wind farm site shall be briefly stated.

7.3.2 The seabed terrain of the planned wind farm site shall be described.

7.3.3 The engineering geological conditions and main engineering geological problems of the planned wind farm site shall be stated.

7.4 Transportation and Construction

7.4.1 The transportation and construction conditions in the region where the planned wind farm is located shall be briefly stated.

7.4.2 The transportation and construction conditions include the water depth, ports, navigation, roads, railways, etc.

8 Planned Installed Capacity

8.0.1 A representative wind farm should be selected as the typical wind farm according to the site-specific characteristics, including the wind energy resource, marine environment, engineering geology, site area, water depth, offshore distance, site geometry, etc.

8.0.2 The wind turbine type and hub height of the typical wind farm shall be preliminarily determined according to the state-of-the-art and development trend of wind turbine manufacture, taking into account the wind regime characteristics, the marine environment, and the transportation, construction and installation conditions.

8.0.3 The characteristics of the planned wind farm site shall be analyzed. Based on the analysis, the principles for arranging the wind turbines of the typical wind farm shall be proposed, the wind turbines shall be preliminarily arranged, and the coefficient of installed capacity shall be proposed following the principles of utilizing marine in an economical and concentrated manner.

8.0.4 The installed capacity of each planned wind farm shall be estimated by using the coefficient of installed capacity of the typical wind farm.

8.0.5 The annual energy output of the planned wind farm to the power grid shall be estimated, and the equivalent annual full-load hours and the capacity coefficient shall be proposed.

9 Preliminary Power System Interconnection Scheme

9.0.1 The current status and development plan of the power system in the region where the planned wind farm is located, including the power mix, power grid structure, and power load and its characteristics, shall be briefly stated.

9.0.2 The capacity of the power grid to accommodate the wind power shall be analyzed and the power consumption market shall be proposed based on site distribution, planned capacity and output characteristics of the wind farms, taking into account the power mix, grid structure, load characteristics, peaking and energy storage measures, etc.

9.0.3 The landing scheme and the power collection and transmission scheme of the planned wind farms and the suggestion for grid connection shall be proposed.

9.0.4 The interconnection pattern between the planned wind farm and the electric power system, the transmission voltage level, and the location and size of the step-up substation shall be proposed, and the geographical connection diagram shall be plotted.

10 Planning Scheme of Typical Wind Farm

10.0.1 For each type of wind farms, a representative site shall be selected; the wind turbine type shall be preliminarily proposed; the layout of wind turbines, cables, step-up substation and control center, construction site(s), etc. shall be made; and the general layout of the typical wind farm shall be proposed.

10.0.2 Based on the construction conditions of the typical wind farm and through preliminary analysis and calculation, the foundation type of the wind turbine suitable for the typical wind farm shall be proposed.

10.0.3 The main electrical connection pattern of the step-up substation, the type selection of main equipment and the layout scheme shall be preliminarily proposed based on the construction conditions and installed capacity of the typical wind farm.

10.0.4 The structure and foundation type of the step-up substation of the typical wind farm shall be proposed based on the layout scheme and construction conditions.

10.0.5 The layout scheme of the onshore control center of the typical wind farm shall be proposed, indicating the floor area, structure and foundation type of each building.

10.0.6 The power collection scheme shall be proposed through preliminary analysis based on the wind turbine layout and the step-up substation location of the typical wind farm; and the preliminary power transmission scheme shall be proposed based on the installed capacity and geographical location of and grid conditions nearby the typical wind farm.

10.0.7 The meteorology, marine hydrology and geology and the transportation, construction and installation conditions of the typical wind farm shall be outlined, the construction scheme of main works shall be proposed, and the marine area to be utilized by the wind farm shall be estimated.

10.0.8 The list of main equipment and materials of the typical wind farm and the work quantities of the wind turbine foundation, step-up substation and control center shall be proposed.

11 Preliminary Environmental Impact Assessment

11.0.1 Data on the current status of the environment in the planned area shall be collected, preliminary field investigations shall be conducted, and the main environmental elements shall be analyzed, identified and screened out.

11.0.2 A preliminary prediction and assessment on the impacts of the wind farm on the main environmental elements in the planned area shall be conducted, and preliminary measures against the main adverse impacts shall be proposed.

11.0.3 A general environmental impact assessment shall be conducted.

12 Rough Investment Estimation

12.0.1 The principles and basis for preparing the rough investment estimate shall be stated.

12.0.2 The total static investment and the investment by year shall be roughly estimated for the typical wind farm.

12.0.3 The total static investment and the investment by year shall be proposed for each planned wind farm based on the construction conditions, taking into account the rough investment estimate of the typical wind farm.

13 Preliminary Analysis of Benefits

13.0.1 The basis and main parameters for financial evaluation shall be briefly described, a preliminary financial evaluation shall be conducted for each planned wind farm, and the main economic benefit indicators shall be proposed. The economics of the offshore wind farms in the planned area shall be analyzed based on technological and industrial development.

13.0.2 The environmental and social benefits from the planned wind farms shall be preliminarily analyzed.

14 Planning Objectives and Development Priorities

14.0.1 The planning objectives for short term, medium term and long term shall be proposed for the planned area according to the national energy development plan and the local economic and energy development requirements, taking into account the wind farm construction conditions and the grid accommodation capacity.

14.0.2 A techno-economic comparison table of the planned wind farms shall be worked out according to the planning objectives, taking into account the preliminary works, construction conditions, grid connection conditions and accommodation capacity, techno-economic indicators, regional economic development, etc. Through comprehensive comparison, the development priorities for the planned wind farms shall be proposed.

15 Conclusions and Suggestions

15.0.1 The planning outcomes shall be summarized, and the main conclusions shall be given.

15.0.2 Based on the planning objectives and development priorities, the suggestions on the preliminary works for offshore wind farms shall be given and should include the observation scheme for wind energy resources and marine hydrology in the planned area.

15.0.3 The critical factors that might restrict or influence the development and construction of the planned wind farms and suggestions on solutions should be proposed according to the local economic development, environmental impacts and grid accommodation capacity.

15.0.4 The suggestions on policies for offshore wind power development and construction and supporting measures should be given.

Appendix A Contents of Planning Report for Offshore Wind Power Project

1 General

    1.1 Background

    1.2 Overview of the Region

    1.3 Overview of Preliminary Works

    1.4 Outcomes of Offshore Wind Power Planning

2 Planning Principles and Basis

    2.1 Guidelines and Principles

    2.2 Basis

    2.3 Target Year

    2.4 Scope of Planning

3 Siting

    3.1 General Distribution of Wind Energy Resources

    3.2 Principles

    3.3 Analysis of Influence Factors

    3.4 Outcomes

4 Project Construction Conditions

    4.1 Wind Energy Resources

    4.2 Marine Hydrology

    4.3 Engineering Geology

    4.4 Transportation and Construction

5 Planned Installed Capacity

    5.1 Selection of Typical Wind Farm

    5.2 Layout of Typical Wind Farm and Planned Installed Capacity of Wind Farms

    5.3 Estimation of Annual On-Grid Energy of Wind Farms

6 Preliminary Power System Interconnection Scheme

    6.1 Overview of the Power System

6.2 Preliminary Analysis of Power Consumption in the Electricity Market

6.3 Preliminary Power System Interconnection Scheme

7 Planning Scheme of Typical Wind Farm

    7.1 General Layout

    7.2 Electrical

    7.3 Civil Works

    7.4 Construction Planning

    7.5 Main Quantities

8 Preliminary Environmental Impact Assessment

    8.1 Overview of the Environment

    8.2 Preliminary Analysis and Assessment on Environmental Impacts

    8.3 Conclusions and Suggestions

9 Rough Investment Estimation

    9.1 Principles and Basis

    9.2 Investment Analysis of Typical Wind Farm

    9.3 Rough Investment Estimation of Planned Wind Farms

10 Preliminary Analysis of Benefits

    10.1 Preliminary Financial Benefit Evaluation

    10.2 Preliminary Analysis of Environmental and Social Benefits

11 Planning Objectives and Development Priorities

    11.1 Planning Objectives

    11.2 Development Priorities

12 Conclusions and Suggestions

    12.1 Conclusions

    12.2 Suggestions

13 Attached Drawings

    13.1 Administrative Division of the Planned Area

    13.2 Distribution of Wind Energy Elements in the Planned Area

    13.3 Geographic Connection of the Planned Wind Farms to the Power

System
13.4 Land Transportation Layout of the Planned Area
13.5 Layout of Main Ports and Shipping Routes in the Planned Area
13.6 Marine Function Zoning of the Planned Area
13.7 Layout of Planned Wind Farms
13.8 Development Time Sequence of Planned Wind Farms

Explanation of Wording in This Specification

1. Words used for different degrees of strictness are explained as follows in order to mark the differences in executing the requirements in this specification:

    1) Words denoting a very strict or mandatory requirement:

       "Must" is used for affirmation; "must not" for negation.

    2) Words denoting a strict requirement under normal conditions:

       "Shall" is used for affirmation; "shall not" for negation.

    3) Words denoting a permission of a slight choice or an indication of the most suitable choice when conditions permit:

       "Should" is used for affirmation; "should not" for negation.

    4) "May" is used to express the option available, sometimes with the conditional permit.

2. "Shall meet the requirements of ..." or "shall comply with ..." is used in this specification to indicate that it is necessary to comply with the requirements stipulated in other relative standards and codes.